Plumbing Answers
101 & *Tips*

By
Sherman Turner

Copyright © 2020 Sherman Turner
All rights reserved. No part of this document may be reproduced or transmitted in any form or by any means, electronic, mechanical, photocopying, recording, or otherwise, without prior written permission of Sherman Turner.

The Table of Contents

About The Book v

Chapter #1 Toilets - 101 1

Chapter #2 Water - 101 6

Chapter #3 Water Tanks - 101 10

Chapter #4 Sewers - 101 13

Chapter #5 Sump Pumps - 101 19

Chapter #6 Plumbing Answers - 101 22

Chapter #7 Plumbing Answers 101 & Tips 43

Chapter #8 Plumbing Estimating - 101 56

About The Author 81

Plumbing Answers
101 & Tips

About The Book

The <u>Author</u> and <u>Master / Plumber</u> ***Sherman Turner*** realize that in today's tough economy many people are wanting to know how to up keep their homes and repairs at the same time while trying to increase savings.

Therefore, this new guidance book ***"Plumbing Answers 101 & Tips"*** will and should enlighten your efforts towards increased savings. The information in this book is designed to help make cost saving decisions. This book is not a do-it-yourself (DIY) book.

<u>Author</u> and <u>Master / Plumber</u> expertise as a retired small business persons information, will benefit small businesses also. Many homeowners and small businesses are getting ripped-off by unscrupulous large Contractors. This book may help you get better informed and increase your home savings.

This book by <u>Master / Plumber</u> specializing estimating will give new informational plumbing charts, so you can be better informed. Our detailed answers to your many questions, will be priceless. Plus, the charts help small businesses be more competitive to win more jobs.

"Plumbing Answers 101 & Tips" is the first book of its kind with many dynamic answers helping to reduce plumbing rip-offs. Thank you for getting my book. Read on and start saving thousands of dollars.

Chapter #1

Toilets - 101

Question: Should a toilet closet flange fit into or over the PVC drain pipe?

They make both kinds. Just make sure it fits snug before you glue it. There are several types a 3x4 fits over 3" and inside 4" but there are ones that will go over 4" and inside 3" also.

Question: Why would a toilet bowl suddenly go dry?

A toilet like any fixture can lose its seal (water). By negative pressure or positive pressure or back siphonage. All are vent related except capillarity action.

The toilet is not vented properly and the action and turbulence of other plumbing in the building is causing the water to be sucked out of the bowl. There is usually a sink in the same bathroom, check to see if it has vent piping or is properly vented?

Plumbing Answers 101 & Tips

Question: What is the best way to move the toilet flange a couple of feet and is there an adapter to go from cast iron to PVC?

Cut the toilet line down at the first 90 from the flange and relocate so that your toilet is still vented from its original vent. If it is dry vented, you need to move the toilet and the vent.

A No-Hub Band or Fernco Coupling is the preferred transition from cast iron to PVC plastic.

Question: What is the correct type seal for a toilet flange? Why should the toilet have this type seal?

Always use deep seal wax gasket. That type seals right inside and on top of the toilet flange, preventing sewer gases from entering the house.

Picture shows the correct placement of the plumbing deep seal wax gasket.

Author SL Turner

Note: Plumbers or handymen who don't seal toilet flange properly causes water leaks resulting in water damages, in thousands of dollars.

Plumbing Answers 101 & Tips

Question: What is the correct wax seal for a toilet flange that is flush with the floor?

Try Ultra Seal type wax gasket, I usually use the thickest wax ring I can find with the plastic funnel thing embedded in it.

Note: The below picture illustrates different parts of the toilet that the Plumber must check is working properly.

Author SL Turner

Please see the most important floor seal at the closet flange. Make sure to check and test that your toilet is fasten securely and does not rock. Did the Plumber seal your toilet fixture space and the floor?

Plumbing Answers 101 & Tips
Chapter #2

Water - 101

Whenever you have a water emergency in the house such as a leak or water causing water damages to property. TURN-OF the water!

Question: How can you increase the water pressure of a faucet?

Usually the aerator is plugged; but, sometimes the rubber inside the stops under the sink wear out and tear away.

Sometimes the supply lines get kinked and need to be replaced. If you get a flex tube to connect to your stop under the sink, you could attach it and blow the water into a bucket to see if you have pressure there. If there is pressure, then you know the problem is above the stops.

Question: What causes a loud knocking sound like a jackhammer in the home plumbing when water is being shut off?

This is called "water hammer" and you need to install "water hammer arrestors" in the piping system where required.

Author SL Turner

Question: What is wrong when pipes make sound as you turn on the water?

It could be you have a loose washer on a faucet. If that is what you mean when turning on the water. Please install new washers and if the faucet become defective, then please buy and install new faucet.

Unscrupulous Contractors will say you need total new piping water replacements. Thanks for asking question, because just saved yourself plumbing bill worth many thousands of dollars.

Question: What causes or how to prevent water piping to rattle, shake and make loud noisy sounds? Does this mean major new water lines and new piping should be installed?

You should have Water Hammer Arrestors installed, same as the types and examples shown below. There are many Companies and Manufactures of Water Hammer Arrestors.

The water piping and water lines should be properly supported with the proper type of insulation as required.

Plumbing Answers 101 & Tips

Author SL Turner

Question: How can I increase water pressure at my kitchen sink Faucet?

When you start having reduced water pressure at your kitchen sink or any sink. This means you may have some blockage in the aerator.

Clean the aerator out or replace with a new one. Then turn your water back on and your water pressure should have increased adequately.

Plumbing Answers 101 & Tips

Chapter #3

Water Tanks - 101

Question: How do you stop a noisy hot water heater from making noise?

You can reduce the noise by draining the Water Heater and removing the lime deposits the best you can.

Most likely you can't reduce the noise if the gas Water Heater and you have hard water. The only way to fix this is to buy a new Water Heater and install a Water Softener.

Author SL Turner

Question: How do you stop pipes and water piping from sweating?

You need to wrap the pipes with insulation. Get insulation that the inside diameter of the insulation is larger than the outside diameter of the piping system.

Note: The wall thickness of the insulation should be at least **"one inch thick."** Make sure to use "duct tape" or "electrical tape" when covering turns in the piping system.

Plumbing Answers 101 & Tips

Question: Why would kitchen sink or washroom sinks have a stink smell?

The only reason I know that the stink smell would come about is that the P-Trap or U-Bend underneath a fixture type sink has lost its trap seal due to evaporation.
Refill the trap seal and pour water into the sinks to have a trap seal.

Question: What causes sink drains to make sucking sounds?

The sucking noise is a good thing, it means that the drain is working properly, and the noise comes from a swirl in the water going counter clockwise and pulls the water. The noise is only air.

Author SL Turner

Chapter #4

Sewers - 101

Question: When roughing in a new toilet what is the correct distance from the center of the drain to the wall?

A minimum of 15 inches off the finished side wall and a minimum of 12 inches of the finished back wall.

However, 18" side clearance is the international standard for wheelchair and assisted toilet maneuvering room. Maybe make more rooms comfortable for larger people.

Plumbing Answers 101 & Tips

Question: How do you make repairs when the Lavatory sink is plugged and not working?

Plumber shown below is using the correct tool for the fixture trap removal. Once the fixture trap is removed and cleaned the Lav sink will drain properly.

Question: What is waterproofing? Where should waterproofing be used?

Use waterproofing sealers and paints which are recommended materials for their usage only. However interior waterproofing sealers and paints will not alleviate all potential sources of basement leaking problem areas.

Author SL Turner

Question: How often should downspouts and gutters be cleaned?

Rain gutters and downspouts should be cleaned regularly. Remove all leaves, dirt and debris at least once or twice a year.

Question: What are "French Drains?" And where should they be used?

"French Drains" are used to channel water out of the building which then would be pumped out of the building through means of a sump pump and sump pump pit.

Question: How often should roofs be inspected?

Roofing systems should be inspected at least once a year.

Question: Who best installs new roofs? A General Contractor or a Roofing Contractor?

A licensed Roofing Contractor who offers a guarantee for their work and materials.

Plumbing Answers 101 & Tips

Question: How long does a roto-rooter job last?

It depends on what's wrong with your sewer. It could take anywhere from 10 minutes to a whole day. Usually blockages in the main occur either from roots or breakages in the pipe.

If the line has been snaked and you still have a problem I would recommend having the sewer, camera for pictures of the interior sewer system piping.

Question: How do you keep roots out of the main sewer line?

The only thing you can really do is snake out the sewer drainage system and get it as clear as possible. Then you can dump copper sulfate down the system.

It probably won't kill all the roots, but it stops them from growing. This works best if you do it in the spring during the growing season. And you should probably do it about twice a year.

Author SL Turner

Another chemical to use is "Root X" because "Cooper Sulfate" is illegal in some states.

Putting any kind of chemicals down your sewer drain line is illegal in most countries around the world; it is considered an environmental hazard. If you are caught by authorities, you may be in serious trouble and the implications will be severe.

Therefore, it may be best to avoid using chemicals and rather rod or snake your sewer line, then cut down brushes or trees suspected to be growing over your sewer. Or replace your sewer line and maintain your sewer line by rodding it every six months.

Plumbing Answers 101 & Tips

Question: What is a plumber's snake used for?

It is a tool they use to shove down your pipes to clean them out if they are clogged and the snake can be either electric powered or by hand. See the example sewer machines below:

Author SL Turner

Chapter #5

Sump Pumps - 101

<u>Question:</u> Every time it rains my basement gets flooded from water under basement floors and walls, what can I do about this situation and problem?

Seems like you may need a sump pump and basin used to manage surface runoff water and from underground water aquifers.

Plumbing Answers 101 & Tips

Question: How do you measure the water evaporation from your swimming pool?

Measuring Water Evaporation First; mark the water level on the wall of the pool. After some time, mark the water level again. It should be lower than the original one.

Measure the area of the pool and then multiply it by the difference of the two water levels. The result is the amount of water evaporated.

It is important to remember to turn off the pool's auto fill for this process. If you keep the auto fill on, then the pool level will not drop much, if any.

Question: What would cause water to seep out from under the toilet bowl onto the floor when the toilet is flushed?

You may have a partial clog somewhere in your pipes, therefore the water is backing up in the first place.

Or the wax seal under the bottom of your toilet (you can't see it because it's UNDER the toilet) is broken.

Author SL Turner

Question: What causes the popping or tinkling sounds as hot metal pipes cool off?

Contraction, though the sounds are usually noisier when pipes are expanding as hot water runs through them. The sound you hear is caused by the expansion and contraction of the metal cause by the heating and cooling.

Question: What would cause the hot water to be rusty and brown?

Sometimes chemicals cause rust to get into the water lines. Or change in water pressure will cause this rust to turn loose and come through into the bathtub or other appliances.

If only the hot water is rusty! It's 100% your hot water tank!

The real reason why you're not seeing it in your cold water is because your cold-water lines run directly to your water fixtures.

Question: Who uses concrete roofs?

It is very common to find commercial and residential buildings having concrete roofing.

Plumbing Answers 101 & Tips

Chapter #6

Plumbing Answers - 101

Faucets and drains are the parts of your plumbing system that are most likely to break down, so therefore you will need repairs more often. Faucets leak or drip and drains get clogged up. When you add these repairs to fixing toilet problems, you have covered almost everything you are likely to encounter.

Toilets: Clogged toilets are the most common plumbing problem can have. If a toilet overflows or flushes sluggishly clear the clog or backup with a plunger or closet auger. If the clog or backup persists, the problem may be in the mains waste and vent system piping.

A recurring puddle of water or water leak recurring on the floor or around the toilet it may be caused by a crack in the base of the toilet. This persistent puddle problem could possibly be from the toilet tank. Also, check all water conditions to the toilet.

Author SL Turner

Toilet stability the toilet fixture shakes or rocks when it is not fastened securely. Check the closet flange and secure all loose connections including replacement of the closet wax ring gasket with nuts and bolts, too.

Water on the floor around toilet - check if the toilet base in the toilet tank is cracked and leaking. Sometimes you may have to insulate the toilet tank to prevent condensation. You may have to tighten the bolts and then check all water conditions.

Plumbing Answers 101 & Tips

Sink drains: every sink has a drain trap and a fixture drain. A sink gets clogs or plug ups by a buildup of soap and hair in the trap or fixture drain line. Remove clogs and plug ups by using a plunger, disconnecting and cleaning the trap or using a hand auger.

Clogged and plugged lavatory sinks can be cleared with a plunger. Remove the pop-up and strainer first, and then plug the overflow in the sink by stuffing a wet rag into it, allowing you to create air pressure with the plunger.

Dump out the debris after removing the trap. You may need a small wire brush when cleaning the trap bend. Reinstall the trap bend and tighten all

Author SL Turner

Faucets eventually just about all faucets develop leaks in drips. Repairs can be accomplished by replacing the mechanical parts inside the faucet body. The main thing is to figure out what kind of faucet you have and know the make of its parts.

If your old faucet continues to leak after repairs are made, then you know that you will have to replace the old faucet with a new faucet.

Water pressure at the spout of the faucet is problematic water pressure seems low water flow is partially blocked. Clean the faucet. This is this is not correct the situation, take a closer look under the sink if the pipe in his old galvanized piping, replace the corroded galvanized pipes with copper or plastic new water piping.

Plumbing Answers 101 & Tips

A dishwasher that is past it's may be inefficient in many ways. It probably was not designed to be very efficient to begin with. However, most significantly, if it no longer cleans effectively, you are probably spending a lot of time and how water pre-rinsing the dishes.

This alone can consume more energy and water in a complete war cycle on your machine. Therefore, even if the old dishwasher still runs, replacing it with an efficient model can be a good investment upgrade.

Plumbing Answers 101 & Tips

Replacing an old, inefficient dishwasher is a straightforward project usually takes just a few hours. The energy savings begins with the first bowl of dishes and continues with every load thereafter. Therefore, we always recommend using a newer model of the appliance.

Food disposers are standard equipment in the modern home and most of us have come to depend on them to macerate our plate leavings, so our crumbs can exit the house along with wastewater from the sink drain.

Author SL Turner

In terms of size and utility hookups, dishwashers are generally quite standard if your old machine is a built-in and your countertops and cabinets are standard sizes.

Most full-size dishwashers will fit right in. Of course, you should always measure the dimensions of the old unit before shopping for the new one to avoid an unpleasant surprise it installation time. Also, be sure to **review the manufacturer's instructions** before starting any work and installations.

Author SL Turner

If your existing disposer needs replacing, you will find that the job is relatively simple, especially if you select a replacement appliance that is the same model as your old one. In that case, you can probably reuse the existing mountain assembly, drain Steve, and drain plumbing.

Most food disposers are classified as are classified as "continuous feed" because they can only operate with an on/off switch on the wall is being actively held down. Let go of the switch and this disposer stops.

Each appliance is a power rating between 1/3 and one HP (horsepower). The more powerful models body down less under heavy loads, and the motors last longer because they do not have to work as hard. They are also less costly too.

Disposers are hard wired to a switch mounted in an electrical box in the wall above the countertop. If your kitchen is not equipped for this, consult a wiring guide or hire an electrician. The actual electrical hookup of the appliance is quite simple, (you only have to join two wires) but hire an electrician if you are not comfortable with the job.

Plumbing Answers 101 & Tips

Icemakers, the most expensive refrigerators come with icemakers as standard equipment, and practically every model features them as an option (a refrigerator with an icemaker usually calls about $100 more).

Most icemakers come either preinstalled or are purchased as an accessory when you buy your **new refrigerator**. An icemaker receives its supply of water for making cubes to through a ¼ inch copper supply line that runs from the icemaker to a water pipe. The supply line runs through a valve in the **refrigerator**, and it is controlled by the solenoid valve that monitors the water supply and sends the water into the icemaker itself, where it is turned into ice cubes.

Author SL Turner

Automatic icemakers are simple to install is long as your refrigerator is icemaker ready. Make sure to buy the correct model for your appliance and try doing careful installation work. Icemakers water supply lines are very common sources we water leaks occur quite often.

Standpipe drains in many houses the washing machine drain holes is followed loosely over the side of the **utility sink** this arrangement is now frowned upon by the building codes.

Instead of hanging it over the **utility sink**, you should install standpipe that allows the **washing machine** to drain directly into the **utility sinks** drain line piping system. A 2-inch standpipe is required by most codes, with a 2-inch P trap. The top of the standpipe should be higher than the highest water level in the **washing machine**.

Plumbing Answers 101 & Tips

Water softeners if your house has hot water running through his pipes, you have got a couple problems. Not only does your water do a poor job of dissolving soap, you also have plenty of scale deposits on your dishes, plumbing fixtures, and the inside of your Hot Water Tank.

Water softeners fix these problems by chemically removing the calcium and magnesium that are responsible for the hard water.

Author SL Turner

Water softeners are installed after the water meter but before the water line branches off to the appliances or fixtures, with one exception. Piping to outside faucets should branch off the mainline before the softener because treating outside water is a waste of money.

Sump pumps, price ranges from about $100-$500 or more, depending on the quality and the features of the sump pump. Below is a picture of a standard sump pump.

First decide between a pedestal and a submersible pump. The standard pump, non-submersible, as shown below:

Plumbing Answers 101 & Tips

Submersible pumps sit in the water a good deal of the time, they have a lifespan from 5 to 10 years. However, most manufacturers offer limited 1 to 5-year warranties. Sump pump is measured by horsepower.

It is better to buy a cast-iron sump pumps, which last longer than plastic or iron types because of corrosion. Make sure the power cord is long enough because electrical extension cords are not to be used on sump pumps.

Always place pavers or bricks underneath the sump pump so that mud, dirt, and grit not plug up the operation of the pump.

Author SL Turner

Using a plunger or a "plumber's plunger" which is about 10 to 12 feet long, to push the obstruction down thru F.A.I. and hose trap, to clear the clogged or plugged up sewer line.

For some hot water down the line and then apply the pressure forced by plunging several times. The pressure from the simple tool can generate pressure to blow out obstructions quite quickly.

Using an **electric snake** if the obstruction is in the form of tree roots, instead of common obstructions like toilet paper rags leavings, and garbage. You need to use a snake tool, which is normally used by plumbers and available in the hardware stores, for a rental fee.

In cases of extreme clogging and plug ups, call a professional plumbing company may use a jet hydro-vac machine.

A clogged sewer is a problem, which needs immediate attention as it can become a nasty nightmare if not taking care of in time.

Plumbing Answers 101 & Tips

The **below** is the type electric sewer machine or **electric snake machine** that you will need:

Author SL Turner

Hot Water Heater problems normally become self-evident. The hot water faucet fails to summon or get hot water from the spout, you see scribbling or puddles near the Hot Water Heater, or the tank emits strange gurgling or pop and sounds coming from the Hot Water Heater.

Hot Water Heaters makes strange noises expanding and contracting metal parts, or more likely, minerals and higher water scale accumulations inside the tank can cause the noises coming from the Hot Water Heater.

Plumbing Answers 101 & Tips

To avoid scale buildups, every few months, open the **drain valve** at the base of the tank, and flush the tank until the rest runs out and you see clearwater.

Most Hot Water Heaters manufacturers recommend draining and flushing your hot water tank once per year or every six months and heart water areas. This helps remove sediment and minerals that collect at the bottom of the tank. Sometimes the sediments chunks may be too large to pass through the drain valve on the tank.

If you hear a boiling sound of water coming from inside your tank this could indicate overheating and very dangerous pressure buildup. You should call a plumber or service professional immediately.

Patch **or** Burst Pipes, if the water pipe freezes and breaks, your priority may be getting it working again, at whatever it takes get the job done. Just remember you are getting a temporary patch job, you still need the pipe or pipes fixed permanently.

Author SL Turner

Kitchen sinks come in many stylish designs now, but you can get basic stainless-steel sinks for around $100-$200. A higher quality stainless steel sink has a higher thicker gauge steel in a higher amount of nickel alloy and surface finish.

The high-quality stainless-steel sink should stay bright for many years and higher quality may cost from $200-$400.

Cast-iron sinks, with enamel finish are very popular. It is possible to chip them, but they have very strong and very durable. Most cast-iron sinks are self-rimming, meaning they have raised lips they rest on top of the countertop, some are available with flush fit rims.

Plumbing Answers 101 & Tips

Types of shower almost nothing is more invigorating than the nice shower. If you have a good showerhead that you use. The type of showerhead you use will have a huge impact on the quality of your shower. There have been many advances in plumbing fixtures, so that your shower heated options are virtually unlimited.

Think showerhead is are fixed directly on the shower wall cannot be removed. However, some fixed chalets are adjustable and can be moved or aimed in different directions, which can be convenient when tall and short people live in the same house share the same shower.

Author SL Turner

<u>Handheld showerheads</u> are connected to a flexible hose that is mounted on the shower wall. With handheld showerheads, you can remove the shower. The attached hose usually allows for greater range of motion, making this the perfect showerhead for bathing pets, clean in the shower stall and hand washing your clothes.

<u>Low-flow shower heads</u> many consumers are turning to low-flow showerheads to reduce energy costs. These showerheads can be fixed or handheld and greatly reduce the amount of water that is sprayed from the nozzle.

Plumbing Answers 101 & Tips

The bathtub uses something called a "trip lever" as the waste system on the tub. There is also a P-trap connected to the trip lever assembly. Most of the time the P-trap will not be clogged or plugged up.

The drain will just be full of hair and can be cleaned out in several ways. How much hair and debris that is wrapped around the tub trip lever. Once you have cleaned all the visible here and debris on the trip lever and drain, apply parts with a light coating of grease.

Take a small hand snake or small electric snake and run 2 or 3 feet of cable to make sure that there are no obstructions in the drain line.

Author SL Turner

Chapter #7

Plumbing Answers 101 & Tips

This very special section of 'Plumber Speaks' details various everyday situations that most homeowners will face when performing plumbing repairs.

By popular request, I have also addressed minor everyday issues, too. If these minor issues are not addressed, they may become major issues, which can result in a loss of hundreds even thousands of dollars in damage.

Knowing and remembering these *Special Tips* can save you 10% to 50% on your plumbing bills.

Faucets and Drains

These are the parts of your plumbing system that are most likely to break down and, therefore, will need repairs more often. Faucets leak or drip and drains get clogged up. When you add these repairs to fixing toilet problems, you have covered almost everything.

Plumbing Answers 101 & Tips

Toilets

Clogged toilets are the most common plumbing problem. If a toilet overflows or flushes sluggishly, clear the clog or backup with a plunger or closet auger. If the clog or backup persists, the problem may be in the main waste and vent system piping.

Puddle of Water or WaterLeak

Recurring water leakage on the floor around the toilet may be caused by a crack in the base of the toilet. This could also be coming from the toilet tank. It is best to check all water connections to the toilet.

Toilet Stability

If the toilet fixture shakes or rocks, then it is not fastened securely. Check the closet flange and secure all loose connections. If loose, replace the closet wax ring gasket with bolts and nuts.

Water on the floor around the toilet tank

Check to see if toilet tank is cracked and leaking. Also, check all water connections to the toilet fixture. You may have to tighten the tank bolts and then check all water connections. Sometimes you may have to insulate the toilet tank to prevent water condensation when very cold water enters a heated room.

Author SL Turner

Sink Drains

Every sink has a drain trap and a fixture drain. A sink gets clogged or plugged because of a build-up of soap and hair in the trap or fixture drain line. Remove clogs by using a

plunger or by disconnecting and cleaning the trap or by using a good hand-auger.

Clogs and Plug-ups

Clogs in sinks can be cleared with a plunger. Remove the pop-up drain plug and strainer on the sink first, and then plug off the overflow hole in the sink fixture by stuffing a wet rag into it, allowing air pressure to build. The air pressure while plunging will blow the obstruction free in the piping. Dump out the debris after removing the trap bend. You may need a small wire brush when cleaning the trap bend. Be sure to tighten all slip nuts.

Faucets

Just about all faucets develop leaks and drips. Repairs can be accomplished by replacing the mechanical parts inside the faucet body. The main thing is to figure out which kind and type of faucet you have and know the "make" of its parts.

Plumbing Answers 101 & Tips

If your old faucet continues to leak after repairs are made, then you know you will have to replace the old faucet with a new one.

Water Pressure

It is problematic when the water pressure at a faucet seems low or the water flow is partially blocked. Clean the faucet aerator. If this does not correct the situation, take a closer look under the sink. Check to see if the piping is old galvanized piping; if it is, then replace the corroded galvanized pipes with new copper or plastic type water piping.

Dishwashers

A dishwasher that is past its prime may be inefficient in many ways. It probably was not designed to be very efficient back in those days to begin with. If your dishwasher no longer cleans effectively, you are wasting time and hot water pre-rinsing the dishes. This alone can consume more energy and water than a complete wash cycle on a newer dishwashing machine. Therefore, even if the old dishwasher still runs, replacing it with an efficient new model is a good investment.

In terms of sizing and utility hookups, dishwashers are generally quite standard. If your old machine is a built-in model and your countertops and cabinets are standard size, most full-size dishwashers will fit right in.

Of course, you should always measure the dimensions of the old unit before shopping for a new one to avoid any unpleasant surprise at installation time. Also, be sure to review and follow all manufacturers' work and installation instructions.

Replacing an Old Inefficient Dishwasher.

This is a pretty straightforward project that usually takes just a few hours. Your energy savings begin with the first load of dishes and continue with every load thereafter. That is why I always recommend using a newer model dishwasher appliance.

Food Disposers

These are standard equipment in modern homes and most of us have come to depend on them to macerate our plate leavings so our crumbs can exit the house along with wastewater from the sink drain, for efficient disposal of organic waste.

If your existing disposer needs to be replaced you will find the job is very easy, especially if you select a replacement appliance that is the same model as the old one. Now you can reuse the same existing mounting assembly and drain plumbing.

Each Appliance has a power rating between 1/3 and 1 HP (horsepower). The more powerful models bog down less under heavy use and the motors last longer because they do not have to work as hard. They are more costly too.

Disposers are hardwired to a switch mounted in an electric box in the wall above the countertop. The actual electrical hookup of the appliance is quite simple (you only must join two wires) but hire an electrician, it's best to be safe.

Icemakers

The most expensive refrigerators come with icemakers as standard equipment and practically every model features them as an option. (A refrigerator with an icemaker usually cost about $100 - $200 more than one without.)

Most icemakers come either pre-installed or are purchased as an accessory when you buy your new refrigerator.

An icemaker receives its supply of water for making cubes through a ¼-inch supply line that runs from the icemaker and connects to a water pipe. The supply line runs through a valve in the refrigerator and is controlled by the solenoid valve that monitors the water supply and sends the water into the icemaker itself, where it is turned into ice cubes.

Then the cubes drop down into a bin. As the ice levels rise, this will raise a bail wire that is connected to water shutoff valve. When the bin gets full, the bail wire triggers a mechanism that shuts off the water supply valve. Unfortunately, water supply lines and the connection points are very common sources of water leaks.

Automatic Icemakers

These are simple to install as long as your refrigerator is icemaker ready. Make sure to buy the correct model for your appliance.

Standpipe Drains

In many houses the washing machine drain hose is hung loosely over the side of the utility sink. This type of installation is now not allowed by many buildings and plumbing codes.

Most codes require homeowners to install a standpipe drain that allows the washing machine to drain into the utility sink's drain line piping system. Using a 2-inch standpipe is mostly required by most codes with a 2-inch P-trap. The top of the standpipe should be higher than the highest level in the washing machine.

Sump Pumps

Prices range from about $150 to $600 or more, depending on the quality and features of a sump pump. First, decide between a pedestal and submersible pump. Which one is best for your purpose and job?

A submersible pump remains out of sight, sits in water a good deal of the time, and has a lifespan of ten years. Most manufactures offer a limited one- to five-year guarantee.

A submersible pump is safe and can be an important advantage if the basement is used as a primary living area. This pump is much safer if there are children in the house.

Sump pump

Is measured by horsepower, which ranges from 1/6 to ½ HP. However, more important is the number of gallons per minute (GPM) a pump will move.

This capacity factor of both the bottom of the pump's efficiency and the "head" or "lift" is the vertical distance from the bottom of the sump to the highest point of discharge. Also, note whether the pump is strong enough to pass small solids, such as leaves and twigs.

Switches on Sump Pumps

These come in several types including floats, diaphragms, and mercury switches. It does not really matter which type you choose but be sure it's automatic.

Buy Cast Iron Sump Pumps

Last longer than plastic or iron type pumps because they don't break easily. Make sure the power cord is long enough because electrical extension cords are not to be used on sump pumps.

Sump Pump installations

Always place "pavers" or "bricks" underneath the sump pump so mud, dirt and grind does not plug up the operation of the pump and motor.

Clogged Sewers.

A clogged sewer is a problem that requires immediate attention. There are some simple techniques to clearing a clogged or plugged up sewer line. Chemical cleaners can effectively clear out any organic obstruction by dissolving

Plumbing Answers 101 & Tips

them. The chemicals are generally available in hardware stores. Be sure to follow the manufactures directions carefully.

Use a "plumber's plunger" which is about 10 or 12 feet long, to generate enough pressure to blow-out obstructions quite quickly

Plumbers and Clogged Sewers

Usually use this at the house trap and fresh-air-inlet outside the house. Use an electric snake when and if the obstruction is in the form of tree roots, instead of common obstructions like toilet paper, rags, leaves and garbage. You may then need the use of an electric snake tool.

For extreme cases, call in a professional Plumbing Company who may use a camera to find the clog and then use a Jet Hydrovac Machine to disperse it.

Water Heater:

Makes strange noises. When the hot water tank is heating water, the expanding and contracting metal parts; or more likely, minerals and hard water scale build-ups these accumulations inside the tank. This can cause the noises coming from your water heater.

Author SL Turner

To avoid scale buildup, every few months, open the drain valve at the base of the tank, then flush out the tank until all the rust and scales runs out and you see clear water.

Hot Water Tank manufacturers recommend draining and flushing your water heater tank every six months in hard water areas. This helps remove sediment and minerals that build-up at the tank bottom.

If you hear boiling sounds inside your hot water tank, this could indicate overheating and very dangerous pressure build-up. Call a professional plumber immediately.

Banging, gurgling, or popping sounds are caused by the build-up of hard water sediment heating up and exploding inside the tank.

Types of Showerheads

The type of showerhead you use will have a huge impact on the quality of your shower. There have been many advances in plumbing, so your choice and options for a showerhead are virtually unlimited. A handheld showerhead is connected to a flexible hose that is mounted on the shower wall. With handheld showerheads, you can remove the showerhead and replace it easily.

Plumbing Answers 101 & Tips

The attached hose usually allows for greater range of motion.

Low-flow showerheads reduce energy costs. These showerheads can be fixed or handheld and greatly reduce the amount of water that is sprayed from the nozzle, allowing for less wasted water.

Kitchen Sinks

These come in many stylish designs now. You can get a basic stainless-steel sink for around $110 to $160. A higher-quality stainless steel sink has a higher (thicker) gauge of steel and a higher amount of nickel alloy and surface finish, but with a much higher cost.

Cast Iron Sinks

With an enamel finish, cast iron sinks are very popular. It is possible to chip them, but they are very strong and very durable. Most cast iron sinks are self-rimming, and some are available with flush-fit rims.

Bathtubs

As one of the most used plumbing fixtures in the house, your bathtub is often covered with soap residue. Your bathtub will eventually get plugged up with hair and dirt and its drain will gradually become clogged.

Older Toilets

When water in an old toilet keeps running and filling the toilet tank, try replacing the old fill water valve inside the toilet tank.

What purpose does Vent Piping serve?

In the Plumbing System. The vent piping system should keep sewer gases from entering your house and help your plumbing drainage piping drain properly.

Chapter #8

Plumbing Estimating - 101

The Author, Master / Plumber has written this book specifically benefit **Home-Owners** and **Small Businesses** who face those tough decisions trying to save money, because upkeep and repairs are costly.

This book takes you by the hand and helps explain the **do's** and **don'ts**. Most plumbing books say do this but never say don't do that. Confusing isn't it!

The most important reason why I wrote this book to help fight against unscrupulous Contractors. Who are always ripping-off the elderly and the disabled.

Author SL Turner

Estimating Piping & Fittings Labor

Notice: All labor rates are reference as average, depending on different job material applications. Labor rates should be adjusted accordingly.

Labor = parts of an hour = .10 = 6 minutes
Labor = parts of an hour = .25 = 15 minutes
Labor = parts of an hour = .5 = 30 minutes
Labor = parts of an hour = .75 = 45 minutes
Labor = parts of an hour = 1.0 = 60 minutes

DWV Copper Pipe Size	Pipe Labor	Fitting Labor
1 1/2	0.01	0.1
2	0.01	0.1
3	0.03	0.15
4	0.04	0.25
6	0.06	0.5

Plumbing Answers 101 & Tips

Estimating Piping & Fittings Labor

Notice: All labor rates are reference as average, depending on different job material applications. Labor rates should be adjusted accordingly.

Labor = parts of an hour = .10 = 6 minutes
Labor = parts of an hour = .25 = 15 minutes
Labor = parts of an hour = .5 = 30 minutes
Labor = parts of an hour = .75 = 45 minutes
Labor = parts of an hour = 1.0 = 60 minutes

Copper "L" Pipe Size	Pipe Labor	Fitting Labor
1/2	0.01	0.1
3/4	0.01	0.1
1	0.01	0.1
1 1/4	0.02	0.15
1 1/2	0.02	0.15
2	0.02	0.25
3	0.03	0.35
4	0.04	0.5

Author SL Turner

Estimating Piping & Fittings Labor

Notice: All labor rates are reference as average, depending on different job material applications. Labor rates should be adjusted accordingly.

Labor = parts of an hour = .10 = 6 minutes
Labor = parts of an hour = .25 = 15 minutes
Labor = parts of an hour = .5 = 30 minutes
Labor = parts of an hour = .75 = 45 minutes
Labor = parts of an hour = 1.0 = 60 minutes

PVC #40 Pipe Size	Pipe Labor	Fitting Labor
1/2	0	0.03
3/4	0	0.03
1	0	0.03
1 1/4	0.01	0.03
1 1/2	0.01	0.03
2	0.02	0.03
3	0.03	0.05
4	0.03	0.05

Plumbing Answers 101 & Tips

Estimating Piping & Fittings Labor

Notice: All labor rates are reference as average, depending on different job material applications. Labor rates should be adjusted accordingly.

Labor = parts of an hour = .10 = 6 minutes
Labor = parts of an hour = .25 = 15 minutes
Labor = parts of an hour = .5 = 30 minutes
Labor = parts of an hour = .75 = 45 minutes
Labor = parts of an hour = 1.0 = 60 minutes

PVC Plastic Pipe Size	Pipe Labor	Fitting Labor
1 1/2	0.02	0.05
2	0.02	0.05
3	0.04	0.05
4	0.05	0.05
6	0.07	0.05
8	0.1	0.1
10	0.15	0.1
12	0.2	0.1

Author SL Turner

Estimating Piping & Fittings Labor

<u>Notice:</u> All labor rates are reference as average, depending on different job material applications. Labor rates should be adjusted accordingly.

Labor = parts of an hour = .10 = 6 minutes
Labor = parts of an hour = .25 = 15 minutes
Labor = parts of an hour = .5 = 30 minutes
Labor = parts of an hour = .75 = 45 minutes
Labor = parts of an hour = 1.0 = 60 minutes

Black Steel Pipe Size	Pipe Labor	Fitting Labor
1/2	0.01	0.1
3/4	0.01	0.1
1	0.02	0.1
1 1/4	0.02	0.15
1 1/2	0.02	0.15
2	0.03	0.25
3	0.04	0.35
4	0.06	0.5

Plumbing Answers 101 & Tips

Estimating Piping & Fittings Labor

Notice: All labor rates are reference as average, depending on different job material applications. Labor rates should be adjusted accordingly.

Labor = parts of an hour = .10 = 6 minutes
Labor = parts of an hour = .25 = 15 minutes
Labor = parts of an hour = .5 = 30 minutes
Labor = parts of an hour = .75 = 45 minutes
Labor = parts of an hour = 1.0 = 60 minutes

No-Hub CI Pipe Size	Pipe Labor	Fitting Labor
1 1/2	0.01	0.1
2	0.01	0.1
3	0.02	0.15
4	0.03	0.25
6	0.04	0.35
8	0.06	0.5
10	0.08	0.75
12	0.1	1

Author SL Turner

Estimating Piping & Fittings Labor

<u>Notice:</u> All labor rates are reference as average, depending on different job material applications. Labor rates should be adjusted accordingly.

Labor = parts of an hour = .10 = 6 minutes
Labor = parts of an hour = .25 = 15 minutes
Labor = parts of an hour = .5 = 30 minutes
Labor = parts of an hour = .75 = 45 minutes
Labor = parts of an hour = 1.0 = 60 minutes

Galv. Steel Support Size	Pipe Labor	Fitting Labor
1 1/2	0.01	0.1
2	0.01	0.1
3	0.02	0.15
4	0.03	0.15
6	0.04	0.25
8	0.06	0.25
10	0.08	0.35
12	0.1	0.35

Plumbing Answers 101 & Tips

Estimating Piping & Fittings Labor

Notice: All labor rates are reference as average, depending on different job material applications. Labor rates should be adjusted accordingly.

Labor = parts of an hour = .10 = 6 minutes
Labor = parts of an hour = .25 = 15 minutes
Labor = parts of an hour = .5 = 30 minutes
Labor = parts of an hour = .75 = 45 minutes
Labor = parts of an hour = 1.0 = 60 minutes

Copper Type Support Size	Pipe Labor	Fitting Labor
1 1/2	0.01	0.1
2	0.01	0.1
3	0.02	0.15
4	0.03	0.15
6	0.04	0.25
8	0.06	0.25
10	0.08	0.35
12	0.1	0.35

Author SL Turner

Estimating Piping & Fittings Labor

Notice: All labor rates are reference as average, depending on different job material applications. Labor rates should be adjusted accordingly.

Labor = parts of an hour = .10 = 6 minutes
Labor = parts of an hour = .25 = 15 minutes
Labor = parts of an hour = .5 = 30 minutes
Labor = parts of an hour = .75 = 45 minutes
Labor = parts of an hour = 1.0 = 60 minutes

Steel Hangers Pipe Size	Pipe Labor	Fitting Labor
1/2	0	0.1
3/4	0	0.1
1	0	0.1
1 1/4	0.01	0.15
1 1/2	0.01	0.15
2	0.02	0.15
3	0.03	0.25
4	0.03	0.25

Plumbing Answers 101 & Tips

Estimating Piping & Fittings Labor

<u>Notice</u>: All labor rates are reference as average, depending on different job material applications. Labor rates should be adjusted accordingly.

Labor = parts of an hour = .10 = 6 minutes
Labor = parts of an hour = .25 = 15 minutes
Labor = parts of an hour = .5 = 30 minutes
Labor = parts of an hour = .75 = 45 minutes
Labor = parts of an hour = 1.0 = 60 minutes

Cop. Hangers Pipe Size	Pipe Labor	Fitting Labor
1/2	0	0.1
3/4	0	0.1
1	0	0.1
1 1/4	0.01	0.15
1 1/2	0.01	0.15
2	0.02	0.15
3	0.03	0.25
4	0.03	0.25

Author SL Turner

Estimating Piping & Fittings Labor

Notice: All labor rates are reference as average, depending on different job material applications. Labor rates should be adjusted accordingly.

Labor = parts of an hour = .10 = 6 minutes
Labor = parts of an hour = .25 = 15 minutes
Labor = parts of an hour = .5 = 30 minutes
Labor = parts of an hour = .75 = 45 minutes
Labor = parts of an hour = 1.0 = 60 minutes

Split Ring Galv. Steel	Pipe Labor	Fitting Labor
1/2	0	0.05
3/4	0	0.05
1	0	0.05
1 1/4	0.01	0.1
1 1/2	0.01	0.1
2	0.02	0.1
3	0.03	0.15
4	0.03	0.15

Plumbing Answers 101 & Tips

Estimating Piping & Fittings Labor

<u>Notice</u>: All labor rates are reference as average, depending on different job material applications. Labor rates should be adjusted accordingly.

Labor = parts of an hour = .10 = 6 minutes
Labor = parts of an hour = .25 = 15 minutes
Labor = parts of an hour = .5 = 30 minutes
Labor = parts of an hour = .75 = 45 minutes
Labor = parts of an hour = 1.0 = 60 minutes

Split Ring Copper	Pipe Labor	Fitting Labor
1/2	0	0.05
3/4	0	0.05
1	0	0.05
1 1/4	0.01	0.1
1 1/2	0.01	0.1
2	0.02	0.1
3	0.03	0.15
4	0.03	0.15

Author SL Turner

PEX Pipe & Fittings

JOB =			
PEX Pipe & Fittings			
Item	Description	Count	Totals

Plumbing Answers 101 & Tips
"L" Copper Pipe & Fittings

JOB =

"L" Pipe & Copper Fittings

Item	Description	Count	Totals

Author SL Turner
IPS Black Steel and Fittings

JOB =			
IPS Black Steel and Fittings			
Item	Description	Count	Totals

Plumbing Answers 101 & Tips
Gas Pipe and Fittings

JOB =			
Gas Pipe and Fittings			
Item	**Description**	**Count**	**Totals**

Author SL Turner
Hangers and Miss. Items

JOB =			
Hangers and Miss. Items			
Item	**Description**	**Count**	Totals

Plumbing Answers 101 & Tips
Plumbing Material Takeoff Sheet

JOB =			
Plumbing Materials =			
Type	Description	Count	Totals

Author SL Turner
PVC WATER

JOB =			
PVC WATER PIPE MATERIALS			
Item	Description	Count	Totals

Plumbing Answers 101 & Tips
PVC DRAIN PIPE

JOB =

PVC DRAIN PIPE MATERIALS

Item	Description	Count	Totals

Author SL Turner

Plumbing Fixtures

JOB =			
Plumbing Fixtures			
Item	Description	Count	Totals

Plumbing Answers 101 & Tips
Plumbing Equipment

JOB =			
Plumbing Equipment =			
Type	**Description**	**Count**	Totals

Author SL Turner
Sub-Contractors Takeoff Sheet

JOB =		
Sub-Contractors =		
System	**Description**	Totals
1		
2		
3		
4		
5		
6		

Plumbing Answers 101 & Tips
Plumbing Job Summary

JOB =			
Summary =			
Item	Work or Job Description	Cost	Totals

Author SL Turner

About The Author

Sherman Turner (1941- now) born in Buffalo, NY to Afro-American parents living in USA. Turner became Buffalo's first Afro-American license Master / Plumber in city and in Plumbers Union #22.

Sherman Turner is a *Master / Plumber* and an Author. He has written other helpful books on plumbing and here is list of many books you can search on Amazon.com for the below listed Plumbing and Estimating books.

Newest books are (**Plumb Pro #1), (Plumb Guide #2)** and **(Plumbing Answers 101 & Tips)** which can be found and listed on Amazon.com. **Other new books** are (Plumber Speaks), (Plumbing 101 Answers), (New Plumbing Guide) and (Plumbing Estimator).

Plumbing Answers 101 & Tips

www.ingramcontent.com/pod-product-compliance
Lightning Source LLC
Chambersburg PA
CBHW070811220526
45466CB00002B/629